Jurassic Fossils
~a brief guide~

Robert Westwood

Inspiring Places Publishing
2 Down Lodge Close
Alderholt
Fordingbridge
SP6 3JA
ISBN 978-0-9564104-9-8
©Robert Westwood 2013
Reprinted 2019, 2021 and 2023

Contains Ordnance Survey data © Crown copyright and database right (2011)

Acknowledgements

Thanks to Charmouth Fossils (underneath the Heritage Centre at Charmouth) and to Lyme Fossil Shop (opposite museum) for allowing me to look round and photograph their collections. Thanks also to the Heritage Centre for their advice.

Particular thanks go to Lyme Regis Town Museum. The director, David Tucker, kindly invited me to photograph as many specimens as I wanted and resident palaeontologist Chris Andrew provided invaluable support whilst I was doing so - not least in being able to open some of the cabinets! Their knowledge is impressive and their enthusiasm refreshing and infectious; I cannot stress strongly enough that this is a great place to start learning about fossils and where to find them. Thanks also to Chris Andrew and Ben Brooks for reading and commenting on the draft.

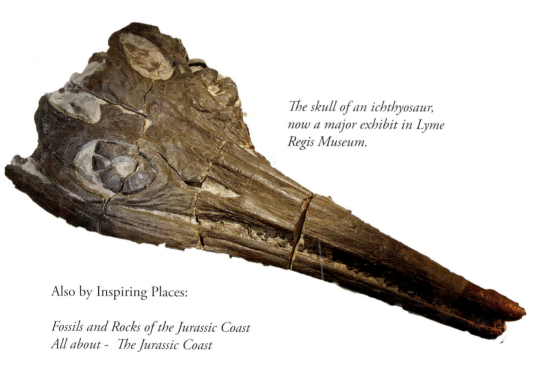

The skull of an ichthyosaur, now a major exhibit in Lyme Regis Museum.

Also by Inspiring Places:

Fossils and Rocks of the Jurassic Coast
All about - The Jurassic Coast

Contents

Introduction	4
Collecting Fossils	6
Places to see, find and learn about fossils	7
The Major Fossil Groups of the Jurassic Coast	8
Belemnites	9
How fossils form	9
Ammonites	10
Bivalves	12
Brachiopods and Gastropods	12
Echinoids	13
Crinoids	14
Brittle Stars	15
Reptiles	16
Dinosaurs	19
Fish	20
Trace Fossils	22
Plants and Trees	23
What to See amd Where to See It	24
East Devon	26
Lyme Regis and Charmouth	28
Seatown and Eype's Mouth	30
Isle of Portland	31
Lulworth, Worbarrow Bay and Dorset County Museum	32
Kimmeridge	34
Isle of Purbeck	36
Classification of Animals and Fossil Code	40
Ownership of Fossils	41

Introduction

The rocks on the Jurassic Coast range from about 250 million to around 65 million years old. Generally, the older rocks are found in the west and they gradually get younger as you travel eastwards. These rocks are from what is called the Mesozoic Era, which is divided into three periods, the Triassic, Jurassic and Cretaceous. It is important to remember that units of geological time are defined by fossils; for example rocks from the Jurassic Period have particular fossils present in them. Some of these creatures only existed for a certain length of time before becoming extinct, thus they are useful time markers. Imagine geologists millions of years from now studying rocks which formed in our time; they will find a collection of fossils that represent life on the planet today. As life on earth evolved so different species appeared and became extinct; the junctions between different periods of geological time represent relatively large changes in life on the planet, and since an "era" is the second most major unit of geological time, the junctions between eras represent very fundamental changes in life.

At the beginning and end of the Mesozoic Era were two "mass extinctions", when over 75% of species died out. Clearly, these are unusual events and causes that have been suggested include huge volcanic outpourings and asteroid impacts. The rocks of the Jurassic Coast thus contain fossils from a time when life was recovering from a global catastrophe of some sort; they document how life recovered and flourished, only to be dealt another devastating blow 65 million years ago. During this time the world was very different from the one we know – weird and wonderful creatures inhabited the warm oceans and the land. The fossils we find in the rocks tell us much

You can see this impressive ammonite near the shore at Winspit on the Isle of Purbeck.

about this incredible history, not only about the creatures and plants that inhabited the planet but also about the environments in which they lived. With a little help it is something everyone can appreciate.

    How do we know how old fossils really are? You might see it stated that an ammonite or dinosaur bone is 100 million years old, but how can we know this? There is nothing about fossils that tells us how old they are. All we can work out from the fossils and the layers they are found in is their relative age, that is that some fossils are older or younger than others. In other words, we can build up a sequence. If you look at the layers of rocks at, say, Kimmeridge the higher the layer the younger it is. We can often follow particular layers to other locations and so continue to work out relative ages. To know how old a rock is in years we need a technique called radioisotope dating. Some rocks, mainly those formed from a molten state (e.g. volcanic), contain radioactive minerals which decay at a known, fixed rate. The amount of decay can be measured and an age for the formation of the rock calculated. This has been done to many rocks around the world which have then been correlated or matched up with other rock layers. By this combination of science and clever detective work we now know the age of many rock layers around the world, and certainly those along the Jurassic Coast.

    I think it is important to appreciate that when geologists talk about the age of fossils they are not simply guessing or picking figures out of the blue; many years of work and study have gone into this and the values are agreed by scientists the world over.

Collecting Fossils

Many people come to the Jurassic Coast looking for fossils. It it is not just fun to collect fossils but also, I believe, to learn about the development and evolution of plants and animals, to visit the many museums which have wonderful collections and to photograph specimens so that they are left for others to see.

Nevertheless, there are many who do want to experience the wonder of discovering the remains of a creature that lived and died 150 million years ago. If fossil collecting is what you want to do then there is really one locality you should head for, the stretch of coast by the neighbouring towns of Charmouth and Lyme Regis. Frequent landslides expose new fossils all the time, and if they are not collected they will only be washed away by the sea. Winter is the best time to collect as storms and rain are constantly eroding more fossils on to the beaches. Needless to say, unstable cliffs are dangerous and you should not go near the bottom of the cliffs, collecting should only be done in the loose, fallen material. Find out the times of the tides before you go and take care not to be cut off. Low tide is the best time to look for specimens. Please follow all local safety advice and if you use a hammer, please wear protective goggles.

On the beaches of Lyme Regis and Charmouth there are always people searching the shingle for a perfectly preserved specimen. If you are a beginner it is possible to spend a lot of time looking and have little to show for it. Don't be put off - the solution is to ask the experts. Guided fossil hunting walks are regularly and frequently organised by **Charmouth Heritage Centre** and **Lyme Regis Museum**. They are run by experts who know where and how to look, and have the knowledge and enthusiasm to inform and inspire. Finally, why not take time to document your finds? Record where they were found and what other fossils were present. You can find out the names at a later date.

Picture: A belemnite and an ammonite lying together. (Courtesy Lyme Regis Museum.)

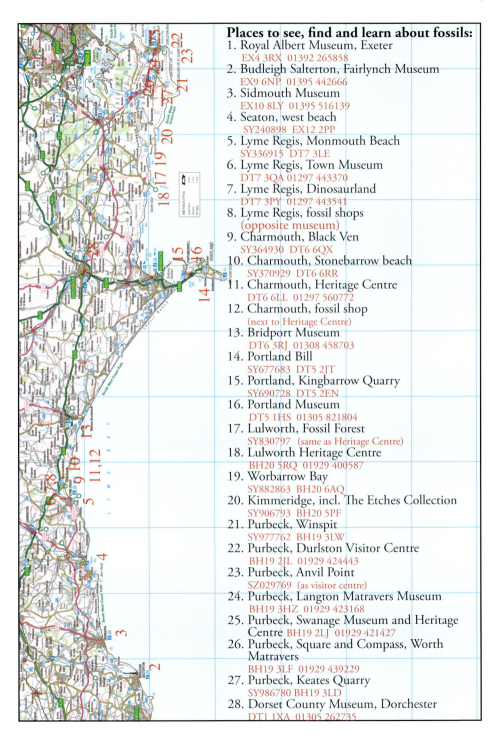

Places to see, find and learn about fossils:
1. Royal Albert Museum, Exeter
   EX4 3RX 01392 265858
2. Budleigh Salterton, Fairlynch Museum
   EX9 6NP 01395 442666
3. Sidmouth Museum
   EX10 8LY 01395 516139
4. Seaton, west beach
   SY240898 EX12 2PP
5. Lyme Regis, Monmouth Beach
   SY336915 DT7 3LE
6. Lyme Regis, Town Museum
   DT7 3QA 01297 443370
7. Lyme Regis, Dinosaurland
   DT7 3PY 01297 443541
8. Lyme Regis, fossil shops
   (opposite museum)
9. Charmouth, Black Ven
   SY364930 DT6 6QX
10. Charmouth, Stonebarrow beach
    SY370929 DT6 6RR
11. Charmouth, Heritage Centre
    DT6 6LL 01297 560772
12. Charmouth, fossil shop
    (next to Heritage Centre)
13. Bridport Museum
    DT6 3RJ 01308 458703
14. Portland Bill
    SY677683 DT5 2JT
15. Portland, Kingbarrow Quarry
    SY690728 DT5 2EN
16. Portland Museum
    DT5 1HS 01305 821804
17. Lulworth, Fossil Forest
    SY830797 (same as Heritage Centre)
18. Lulworth Heritage Centre
    BH20 5RQ 01929 400587
19. Worbarrow Bay
    SY882863 BH20 6AQ
20. Kimmeridge, incl. The Etches Collection
    SY906793 BH20 5PF
21. Purbeck, Winspit
    SY977762 BH19 3LW
22. Purbeck, Durlston Visitor Centre
    BH19 2JL 01929 424443
23. Purbeck, Anvil Point
    SZ029769 (as visitor centre)
24. Purbeck, Langton Matravers Museum
    BH19 3HZ 01929 423168
25. Purbeck, Swanage Museum and Heritage Centre BH19 2LJ 01929 421427
26. Purbeck, Square and Compass, Worth Matravers
    BH19 3LF 01929 439229
27. Purbeck, Keates Quarry
    SY986780 BH19 3LD
28. Dorset County Museum, Dorchester
    DT1 1XA 01305 262735

The Major Fossil Groups of the Jurassic Coast

During the Mesozoic Era (the Triassic, Jurassic and Cretaceous periods) the part of the world we now know as East Devon and Dorset was either covered by a warm, shallow sea or was part of a continent bordering an ocean. These various environments were teeming with life but only very few of the many classes of plants and animals have left abundant fossils. Most fossils are found in rocks that were deposited in the shallow seas that surround continents; conditions on land are much less favourable to the formation of fossils. The soft parts of plants and animals almost always decay after death so it is only usually hard parts that are preserved. The following pages describe the main fossil groups that are found in the rocks of the Jurassic Coast. There are many other types of fossil more rarely found but it is beyond the scope of this book to include them.

Below is a table showing the geological time scale. Note that most of geological time is defined as "Pre-Cambrian"; easily recognizable fossils are not found in rocks before the Cambrian Period.

Era	Period		Age[my]
Cenozoic	Quaternary	Neogene	25
		Palaeogene	65
Mesozoic		Cretaceous	145
		Jurassic	200
		Triassic	250
Palaeozoic		Permian	300
		Carboniferous	360
		Devonian	415
		Silurian	445
		Ordovician	490
		Cambrian	550
Pre-Cambrian	(my = million years)		4600

Belemnites

Along with the ammonites, belemnites are one of the commonest fossils found in the rocks of the Jurassic Coast. Like the ammonites they were an "order" of the cephalopods, marine molluscs characterised by bilateral body symmetry, prominent heads and long tentacles. They had an internal skeleton, the main part of which was the guard or rostrum – this is the "bullet" which is commonly all that is found. The soft body had ten tentacles with small hooks for grasping prey. They were free swimming and, like squids, had ink sacs and large eyes. Belemnites first appeared in the early Jurassic and survived until the end of the Cretaceous.

How Fossils Form

Fossils are almost always found in sedimentary rocks, rocks that have been deposited by the action of water, ice or wind. Additionally, sediments that have been deposited in the shallow "shelf" seas that surround the continents contain the most fossils. It is these seas that, since the Cambrian Period, have supported abundant life, and animals that die here sink to the bottom and can quickly be covered by soft sediment. In contrast animals that die on land usually leave no trace. Hence the fossil record is skewed in favour of marine creatures. Dinosaur fossils are relatively rare although many have been found worldwide. They may have been preserved in the sediment from shallow lakes in which they waded or in seasonal lakes whose water covered previously dry land.

Animals can be fossilised in a number of different ways but it is important to remember that it is almost always the hard parts that become fossilised while soft tissues decompose. However, soft tissues can leave impressions in soft sediments which, given the right environmental conditions, can be preserved. The hard parts themselves can sometimes be preserved – thus we find original shell fragments or bone lying trapped in sedimentary layers. Sometimes the original material is dissolved away leaving a cast or mould of the creature. Occasionally the original material may be chemically replaced during the formation of the sediment (typically when it is being compacted and turned to stone) resulting in a cast of the animal. The wonderful ammonite fossils made of fools' gold (iron pyrite) fall into this category.

Ammonites

Ammonites are marine molluscs from a class commonly known as cephalopods. They are closely related to modern octopuses, squid and cuttlefish but they looked very much like a more distant modern relative, the Nautilus. These spiral shaped creatures swam upright in the ocean, their soft bodies with tentacles residing in the end chamber of their shells. With keen eyes and a well developed nervous system, they were effective predators, seizing their prey with tentacles before devouring it with powerful jaws. Ammonites must have dominated the Mesozoic seas, cruising the shallow, tropical waters of the newly formed oceans and regulating their depth by pumping gas in and out of the chambers of their shells. They ranged in size from a few millimetres across to several metres.

Ammonites first appeared in the Devonian Period around 400 million years ago. They are extremely useful to geologists because of the manner and speed in which they evolved (see Zone fossils opposite). Ammonites evolved particularly quickly, with new species appearing and disappearing throughout the Mesozoic. Some of the ways in which they evolved are also easy to spot; the outside of their shells became more complex and ornamented and the divisions between the chambers of the shell became more and more intricate. This has led to ammonites becoming "zone fossils" for the Mesozoic Era, with particular species indicating that a rock layer is a certain age. For example, the large ammonite pictured on page 5 (the aptly named Titanites) is a zone fossil for the Portland Stone.

Above: Note the lines marking the divisions between the chambers in the ammonite's shell. You can see these are rather more complex than the ones in the photograph on the previous page. (Courtesy Lyme Fossil Shop.)

Zone fossils

The fossil record helps us understand the evolution of plants and animals. Although far from complete, there is enough evidence to show how plant and animal groups have evolved different species and been able to adapt to changing environments. The constant upheavals caused by the movement of the tectonic plates has undoubtedly been a driving factor in the evolution of species. Perhaps partly as a response to changing environments, some animal groups have evolved faster than others, with different species that evolved rapidly and died out. When they leave fossils, such animal groups are of great use to the geologist as they provide valuable time markers. If a species only existed for a relatively short time, finding a fossil example effectively dates the rock in which it was contained. If the creature was perhaps free swimming and was able to spread around the world, such a creature might be useful in correlating or matching up rock strata from different regions. Fossils like these may be used as "zone fossils", they are said to define a particular age. Ammonites are particularly useful for this as they not only evolved very rapidly (as described opposite) but they were, of course, also free swimming, meaning that the distribution of fossil ammonites is quite widespread. Echinoids (sea-urchins) are also used as zone fossils.

Bivalves

Bivalves are another class of the molluscs. They are found all around the world and it is a rare beach that does not have bivalve shells on it. First seen in the Cambrian Period nearly 600 million years ago, they evolved into many different species with varied shapes and sizes. There are still many species around today: giant clams, oysters and mussels are all bivalves and all are based on bilaterally symmetric shells. Many have shells that are mirror images. Some species have been around for millions of years and bivalves are not generally as useful for indicating the relative age of rock strata as the ammonites.

Bivalves siphoned water in and out of their shells, extracting small food particles in the process. Some burrowed in the sand on the sea bed, some attached themselves to rocks and others were free swimming. There were species that lived in fresh or brackish water and these have been useful in indicating that Purbeck limestones were deposited in shallow, coastal lagoons.

Above: Gryphaea arcuata or "Devil's toenail"; this bivalve rested in the mud of the Jurassic and Cretaceous seas, its strongly curved shell keeping the aperture through which it fed clear of the sediment. They are particularly common in places on the Jurassic Coast.

Oysters (see page 31)
Oysters are bivalves that comprise a number of groups which include some that are edible and some that produce pearls. They live in the intertidal region, or in very shallow water close to the shore, attached to the bottom from where they filter food out of the water. They are therefore important indicators of environment.

Brachiopods and Gastropods

A fossil gastropod.

Another group of fossil shells that resemble bivalves are brachiopods. Although they may look similar these creatures are not molluscs and the two groups are not closely related. They are not so common as bivalves and have a different symmetry to their shells.

Gastropods are snails that are found in both Jurassic and Cretaceous strata. They are easily recognisable by their coiled or spiralled shells. The Purbeck limestones, particularly the Purbeck Marble, contain many fossils of the pond snail Viviparus, often tightly packed together.

Echinoids

Echinoids are commonly known as sea-urchins. They are a group from the much larger echinoderm phylum which comprises marine animals with a typical five fold radial symmetry. Many species have prodigious powers of tissue and organ regeneration. They have a rigid skeleton embedded in their skin made of calcite plates, often sold in seaside souvenir shops. Their internal organs were suspended inside the shell and supported by fluid. Echinoids first appeared in the Ordovician Period and are still around today. Only two lineages survived the Permian extinction but they diversified again in the late Triassic and early Jurassic.

Many inhabited shallow water marine environments and moved around grazing on the sea floor, but they are known from almost every major marine habitat, from the intertidal zone to the ocean depths. Some species were able to burrow in soft sediment. So called "regular" forms have the typical five-fold radial symmetry (note photographs above and below left) with areas resembling the spokes of a wheel. Tube feet projected from these spokes and were involved in respiration, feeding and movement. The mouth of the echinoid was on the underside and the anus in the middle of the spokes on the top. Echinoids possessed a "water vascular system"; effectively

a hydraulic system that pushed water in the tube feet enabling them to be used to move the creature around and to collect food. They also had spines covering their bodies which served a defensive purpose as well as being used in locomotion. Echinoids evolved rapidly in the Mesozoic Era and are important zone fossils of the Cretaceous.

Left and top: Fossil echinoids.

Crinoids

Crinoids are often known as sea lilies, but they are animals not plants, in fact they are echinoderms like the echinoids. A very successful group, they have been around since the early Ordovician and are still prolific today. Most of the fossil crinoids we find were fixed to the sea bed by root like structures on the end of a long segmented stalk. The main parts of the animal were in the cup or "calyx" at the top of the stalk and it is quite rare to find these.

The crinoids in this photograph are wonderful specimens from the Lyme Regis Museum. They clearly show the rarely preserved cups attached to the stalks. Notice how the stalks are segmented – it is these little rings that are most commonly found as the crinoids were often broken up after death by currents on the sea bed. It is thought these particular crinoids lived attached to driftwood from the coastal Jurassic forests. At this time the bottom of the sea was anoxic or depleted in dissolved oxygen.

Brittle Stars

Like echinoids and crinoids, brittle stars are also echinoderms. They, too, had a water vascular system and tube feet. Brittle stars are closely related to starfish and first appeared in the Ordovician Period. They are still widespread today. Like starfish they had a five fold radial symmetry, having five arms joined to a central body disc.

Brittle stars generally scavenged for food on the sea floor, moving particles to their mouth on the underside of the body disc by their tube feet. They could move fairly quickly by moving their arms.

Their five, slender arms had an internal skeleton made out of calcite (calcium carbonate) "ossicles". When they died these usually broke up so it is rare to find a complete brittle star like the one in the photograph. Usually just the tiny cylindrical ossicles or sections of the arms are all that are found.

The famous Starfish Bed at Eype's Mouth contains many specimens of brittle stars, but even here complete specimens are quite rare. Much more common are broken bits from the five arms. It is thought that vast numbers of brittle stars were buried by a relatively sudden influx of sediment in the ancient sea - a consequence of a huge storm. If you are lucky you may find a few bits and pieces in the Starfish Bed (see page 30) but nearby Bridport Museum has some wonderful specimens in its impressive fossil collection.

Left: Brittle star, Lyme Regis Museum.

Reptiles

The Mesozoic Era is sometimes known as "The Age of Reptiles". Reptiles are a class of vertebrates that typically are cold-blooded, lay eggs and have scaly skin. They first appeared over 300 million years ago in the Carboniferous Period. They were descended from earlier amphibians and became able to live on land because they had evolved a hard egg able to withstand dry conditions. After the great extinction at the end of the Permian Period, a surviving group known as the diapsids became dominant, diversifying widely. There are thousands of diapsid species alive today, including crocodiles, lizards, snakes and birds. They are classified according to their pattern of skull openings, arrangements which allowed for the attachment of large jaw muscles.

In the Triassic the diapsids were small to medium sized meat eaters but they evolved into a number of groups that came to dominate the Mesozoic landscape in terms of size and numbers. Marine reptiles such as the plesiosaurs and pliosaurs are diapsids and another group known as the archosaurs included the rhynchosaurs, crocodiles and dinosaurs. (We will look at the dinosaurs later.)

Many important specimens of these fantastic creatures were discovered on the Jurassic Coast, particularly around Lyme Regis, by the renowned Victorian fossil collector Mary Anning. Important finds continue to be unearthed.

Plesiosaurs

First discovered by Mary Anning, these marine reptiles first appeared in the Lower Jurassic. They had broad bodies, short tails and two huge pairs of flippers. They were possibly the largest marine predators of all time with some reaching 15 metres in length. Examples have been found with the remains of belemnites and ammonites in their stomachs.

Pliosaurs

Pliosaurs lived in the Jurassic and Cretaceous. They were essentially short-necked plesiosaurs with large heads and massive jaws and are distant relatives of modern lizards. They ranged in length from 4-15 metres. A huge specimen has been unearthed on the Jurassic Coast by amateur fossil hunter Kevan Sheehan. It has been purchased by Dorset County Council and can be seen at the Dorset County Museum in Dorchester.

Crocodiles

Crocodiles were one of a group of land tetrapods that evolved towards the end of the Triassic Period. The early ones were largely terrestrial and had extensive body armour. They were more abundant and diverse during the Jurassic and Cretaceous than they are today. Crocodiles thrived in the swamps where the Purbeck strata were formed and one of the commonest types of vertebrate fossil to be found there is crocodile teeth; partly due to the fact that crocodiles replace their teeth throughout life. In 2009 Richard Edmonds, then Earth Science Manager of the Jurassic Coast team, uncovered a superb fossil skull of a previously unknown species in the cliffs at Durlston Bay. A replica is on display in the Rock Room at Durlston Country Park.

Ichthyosaurs

Ichthyosaurs were large fish shaped reptiles. They evolved from land reptiles in the Triassic around 245 million years ago and became extinct around 90 million years ago. Thus they were around for most of the Mesozoic Era and were the dominant marine predators during the Jurassic, feasting on ammonites, belemnites and fish. On average they were between 2 and 4 metres long although some were much larger. They were fast swimmers (25 mph has been estimated) and generated power by beating their body and tail from side to side, their front paddles being used to steer. It is thought they could also dive deeply. They were warm blooded, air breathing and gave birth to live young in the water.

Part of an ichthyosaur skeleton, Lyme Regis Museum.

Ichthyosaurs resembled dolphins in their morphology and lifestyle and have often been cited as as an example of convergent evolution. Although the two groups are not closely related, natural selection has led to them developing similar forms and features in response to the similar challenges in their environments.

The first complete ichthyosaur fossil was found near Lyme Regis by Mary and Joseph Anning in 1811. It is now in the Natural History Museum.

Above: The giant skull of an ichthyosaur acquired by Lyme Regis Museum (also on page 2).

Left: Part of the backbone of this huge creature.

The ichthyosaur pictured above and on page 2 forms a major exhibit at Lyme Regis Museum. It was purchased by the museum from local collector Mike Harrison who found it a short distance from the museum. The skull measures over 1.5 metres in length and is the same species as the first complete specimen unearthed by Mary Anning.

The ichthyosaur pictured on page 17 is also on display in the museum and is much smaller. In amongst the broken ribs can be seen rare fossilised traces of soft tissues including the contents of its stomach.

Dinosaurs

The name "dinosaur" comes from Greek words meaning "terrible lizard". However, dinosaurs were not lizards, they evolved from the diapsid reptiles during the Triassic Period and the first ones were relatively small carnivores. Classifying the remains of reptiles as dinosaurs is a complicated business, but palaeontologists have agreed a series of anatomical features that all dinosaurs shared. There were three main groups; the sauropods were herbivores and developed into colossal forms like Brachiosaurus which could be up to 25 metres long; the theropods which were mostly bipedal carnivores such as Tyrannosaurus Rex and the ornithischians which were mostly bipedal herbivores. This last group also developed some very large species, notably Stegosaurus.

Dinosaurs came to dominate the Mesozoic landscape and were undoubtedly the top land predators of the time. They were extremely successful, radiating into thousands of different species and exploiting many different environments. There has long been a debate as to whether dinosaurs were warm or cold blooded. It is thought that some of the smaller predators might have been warm blooded, but a great deal of extra food would have been needed to keep the core temperature of the giant herbivores constant. However, such was their bulk that the large dinosaurs would have kept a fairly steady core temperature anyway, whether or not they were warm blooded.

Dinosaur fossils are rare, not least because they lived on land and their remains would seldom be preserved. However, some were preserved in the soft sediment on the bottom of seasonal lakes or coastal swamps. On the Jurassic Coast we have such sediments; the red Triassic sandstones of east Devon were formed in a desert environment by seasonal lakes and rivers; and the early Cretaceous rocks of Purbeck were formed in coastal swamps and river deltas. In Devon a very good specimen of a rhynchosaur was found in the red sandstone and is on display in the Royal Albert Museum in Exeter. Rhynchosaurs were not dinosaurs, but large diapsid, herbivorous reptiles. The most likely dinosaur fossil you will come across is a footprint in the Purbeck strata of east Dorset. Footprints of large sauropods can be seen at Keates Quarry in Purbeck (see page 38).

Drawing of an Iguanodon.

Fish

Fishes consist of several classes of swimming vertebrates and first evolved in the late Cambrian / early Ordovician. The late Jurassic and early Cretaceous times saw a big radiation in the group of fishes known as teleosts. This is the biggest group today and includes around 23000 species. Their success is probably due to the evolution of an advanced jaw mechanism which enabled them to suck in prey and bite more precisely. Fossil fish are found in many places along the Jurassic Coast and Charmouth in particular is one of the best places in the world to search for Lower Jurassic fishes. However, finding them is still difficult.

The earliest known sharks date from the Ordovician about 420 million years ago but most modern sharks first appeared in a big radiation of species during the early Cretaceous, around 100 million years before the present. Shark skeletons are relatively soft and so they were rarely preserved as fossils; their teeth however are quite common at a number of places along the Jurassic Coast. This is explained by the fact that sharks constantly replace teeth during their life and it has been estimated that some sharks may lose as many as thirty thousand teeth during a lifetime.

Above: Purbeck limestone from Durlston Bay, showing what are probably lots of bits of fish teeth and bone (black specks). Look out for larger pointed sharks' teeth.
Left: A Jurassic fish (by kind permission Swanage Museum and Heritage Centre).

The fossil fish pictured above is a superb example of a Dapedium from the Lower Jurassic. It was found between Lyme Regis and Charmouth in 2009 by Tom Loughlin in a layer of rock known as "Lang's Fish Bed". This is a shale with fibrous calcite crystals that resemble meat - hence it is often known as "shales with beef". He noticed the edge of the fish in a flat block of shale, it was only when the block was split that this beautiful specimen was revealed. Tom was out looking with Lyme Regis Museum's Chris Andrew.

The Dapedium was a deep bodied fish that swam through the water column and ranged from about 9 to 40 centimetres in length. Its skin was covered with thick scales and its skull was armoured with bony plates, especially around the eyes. Its tail was short and strong, perhaps enabling it to change direction quickly. Note the smaller picture showing a close up of its sharp, pointed teeth - the Dapedium was a predator that probably fed on shelled creatures such as small ammonites, bivalves and sea-urchins.

Trace Fossils

Footprints on a sandy beach are short lived; the advancing tide or gusting wind soon erodes them. Imagine, however, walking across the drying bottom of a seasonal desert lake; the mud is soon baked hard and following next winter's floods your hardened footprints would be filled by fresh silt. As the process continues further layers of silt accumulate. Many thousands of years later when the climate has changed again your fossil footprints may be revealed by the erosion of layers above. Trace fossils such as footprints are the remains of animal behaviour. They include tracks and trails, burrows and borings, faecal pellets and even the root penetration structures of plants and trees. It is not often possible to exactly identify what made the traces so they are classified on the type of structure. Trace fossils are often good indicators of the type of environment in which the rocks were deposited. Perhaps the most famous of these fossils are dinosaur footprints which have been found in a number of locations on the Jurassic Coast and can be seen at Keates Quarry in Purbeck (see pages 38,39).

Below: Worm burrows at Winspit on the Isle of Purbeck. The Purbeck limestone was deposited in a shallow coastal lagoon. Worms burrowed in the soft sand and mud on its bottom and these have been preserved as they subsequently filled with sediment. Being harder than the surrounding rock they have been weathered out. Note the bivalve also present.

Plants and Trees

Above: The fossilised remains of an algal burr that grew around a rotting tree stump in a Jurassic coastal swamp. The hole in the middle shows where the tree once was. Inset: Fossilised wood.

The first plants began to colonise the margins of continents in the Ordovician Period around 500-450 million years ago. They were well established by the time the Mesozoic Era started and modern conifers evolved during the late Triassic. In the Jurassic when dry, arid conditions had given way to a warm, humid climate verdant jungles covered much of the landscape. Conifers and tree ferns dominated and the "Fossil Forest" at Lulworth has the remains of largely juniper and cypress trees.

Fossil wood is found at several locations on the Jurassic Coast, places that, during parts of the Jurassic Period, were coastal swamps or river deltas. Wood can be petrified by the infiltration of a mineral solution (calcite or silica) which, as it solidifies, can preserve a great amount of detail of the internal structure.

The face of the planet was changed forever by an event in the middle of the Cretaceous – this was when the angiosperms or flowering plants radiated dramatically. Prior to this event, no colourful flowers would have graced the landscape. Today angiosperms are the most successful plant group.

What to see and where to see it

The ammonite pavement, Monmouth Beach, Lyme Regis.

East Devon

Much of the East Devon coast is formed from red sandstone. These rocks are from the Triassic Period and were deposited in desert conditions; some by seasonal lakes, by rivers or by the wind. They are largely unfossiliferous as plants and animals that lived and died there were unlikely to be preserved as fossils. Nevertheless some important finds have been made, including the skeleton of a rhynchosaur, a large plant eating reptile that is now displayed in the Royal Albert Museum in Exeter.

There is, however, still much to see. Two places in particular have outcrops of fossiliferous rocks. **Seaton**, on its western side has Jurassic and Cretaceous sediments that contain ammonites, echinoids and bivalves. Look in the debris on the beach and in the material fallen out of the cliffs. The charming resort of **Beer** has chalk cliffs which contain echinoids and ammonites. These can be found in fallen blocks but the rock is hard, the cliffs are dangerous and it is not a recommended place for fossil hunting.

Elsewhere the red sandstone cliffs yield interesting information about the sorts of environments where the rocks were formed. Look out for "current bedding" (see photo below); the sloping lines in the sandstones reveal where channels of seasonal rivers have moved, cutting across layers of sand that had previously been deposited. Some of these structures (typically larger) have been formed by shifting desert dunes.

At **Budleigh Salterton** is the famous Pebble Bed. This can be found at the bottom of the cliff just west of the beach. The pebbles were deposited

Current bedding in the cliffs at Sidmouth.

by fast flowing rivers coming from mountains in what is now Brittany. Some, when split, have been found to contain fossils millions of years older than the rock itself – they had previously been fossils in this ancient mountain range! Look closely at the top of the Pebble Bed and you will see very angular pebbles, an indication of wind erosion, showing that the rivers suddenly stopped flowing and desert conditions returned.

Above right: This red pebble is from the top of the Budleigh Salterton Pebble Bed (top picture). Its angular shape shows that desert conditions had returned where previously there had been fast flowing rivers.

The Royal Albert Museum in Exeter has a wonderful fossil collection, including fish, amphibians and plant remains from the cliffs between Exmouth and Sidmouth. It also houses a fossil skeleton of a rhynchosaur, a wide ranging herbivorous reptile from the Triassic that thrived in the desert landscape, feeding on vegetation around the seasonal rivers and streams.

Sidmouth Museum is much smaller than Exeter but makes up for this with charm and friendliness. As well as displays on local history it has an impressive section on the Jurassic Coast including important reptile remains.

The Fairlynch Museum in Budleigh Salterton is housed in a delightful thatched building near the sea front. It has a very good collection of geological specimens and much information about the famous Budleigh Salterton Pebble Beds.

Lyme Regis and Charmouth

Lyme Regis Museum in the heart of the town has a geology gallery with a wonderful collection of fossils, including reptiles, ammonites and amazing fossil crinoids. There is lots of information and the museum also organises regular and frequent fossil walks. See the website for more information, www.lymeregismuseum.co.uk.

The **Dinosaurland Fossil Museum** in Coombe Street has a spectacular collection of local fossils.

There are several fossil shops in the town, **Lyme Fossil Shop** opposite the museum has a particularly fine collection of specimens.

At Charmouth the **Charmouth Heritage and Coast Centre** also has a good collection of fossils and organises fossil walks, rocks and fossil weekends and many other activities. Full information can be found on their website www.charmouth.org.

Just below the Heritage Centre, **Charmouth Fossils** stocks another amazing collection of fossils for sale.

The cliffs at Lyme Regis and Charmouth largely consist of shales, clays and limestones from the early Jurassic around 200 million years ago. These sediments were deposited in a relatively deep sea that teemed with life. On the whole the rock strata dip gently towards the east, a result of later earth movements, meaning that you encounter younger layers as you move eastwards.

    I strongly advise anyone interested in collecting fossils to book a fossil walk with either the museum at Lyme Regis or the Heritage Centre at Charmouth. In Lyme Regis, Monmouth Beach (part of the Axmouth to Lyme Regis Undercliffs National Nature Reserve managed by Natural England) west of the Cobb and Church Cliffs east of the River Lym are the best places to look. In Charmouth the beaches east and west of the Heritage Centre are popular with fossil hunters. Search among the rocks and pebbles on the beach, but please do not go near the base of cliffs and heed any local safety warnings, there is an ever present danger of landslips. (Check websites for South West Coast Path and Natural England.)

    There are many large ammonites typically to be seen in the boulders along the beaches; why not "collect" these with a camera. Don't forget to visit the displays in the museums and Heritage Centre, and to have a look in the fossil shops.

    Finally, although not common, Lyme Regis is the best place in the world for Lower Jurassic insects.

Main picture: The beach east of the Heritage Centre at Charmouth. Look in the debris along the beach.
Left: Monmouth Beach, Lyme Regis has many large ammonites.

Monmouth Beach (car park by beach) east of the Cobb at Lyme Regis is where you will find the "ammonite pavement" (see page 24,25), a ledge of limestone at beach level (only accessible at low tide) which contains hundreds of ammonites, many of them large. This is also a good place to see crinoids, fossil wood and several kinds of trace fossils. By all means look in the debris and pick up small specimens but this is the place to use your camera. **Black Ven** (car park by Heritage Centre) west of the Heritage Centre at Charmouth is another place to look in the fallen material on the beach. Search for ammonites, belemnites, crinoids and reptile bones. Sometimes the harder blocks of limestone contain well preserved ammonites. For these you will need a hammer and chisel, but knowing where to look is key - this is where the Heritage Centre can help!

Stonebarrow (car park as Black Ven) east of the small river at Charmouth is another good place. Here you may be lucky enough to find ammonites made of fools' gold or iron pyrite.

Seatown and Eype's Mouth *Above: The cliffs at Seatown are fossiliferous, at the bottom is the Green Ammonite Mudstone.*

Lower Jurassic rocks are also exposed along the coast between Seatown and Eype's Mouth. The clay which forms much of the lower cliff at Eype is not particularly fossiliferous, but on top of this is a relatively thin band of light coloured sandstone - the famous Starfish Bed which contains the brittle stars. Blocks of this litter the beach at Eype's Mouth and you may find bits of broken "arms" - whole specimens are rare. Lookout too for blocks of a pinkish-brown rock known as the Junction Bed (or Beacon Limestone). This is very fossiliferous and yields many ammonites, although the rock is very hard.

The Isle of Portland

As you might expect, the rock that dominates the Isle of Portland is the Portland Stone. This limestone was deposited in shallow, tropical seas and contains many fossils of marine creatures. The Purbeck series also outcrops here and contains fossils of freshwater bivalves, gastropods and trace fossils. Having said that, Portland is not an ideal location for fossil collecting – many of the ammonites in the Portland Stone need to be hammered out of the rock. Nevertheless, fossils can be seen and are best photographed rather than taken away.

At Portland Bill, on the quarried ledges near the lighthouse can be seen many oysters and other bivalves embedded in the rock. Trace fossils such as burrows are also common. At the northern end of the isle, behind the Purbeck Heights Hotel lies the disused Kingbarrow Quarry. On a flat ledge in the south-west corner of this can be seen a few examples of algal burrs, fossilised algal colonies that formed around a decaying tree stump towards the end of the Jurassic Period. These are similar to those found at the Fossil Forest near Lulworth Cove. Just outside the Portland Heights Hotel is a fine example of a fossilised tree trunk.

Finally, why not take a look in the charming **Portland Museum** near Church Ope Cove. This contains a large collection of local fossils, including a nationally important collection of fossil cycads, primitive trees superficially similar to palms and ferns. There are relatively few species around today.

The Portland limestone, left, is near the tip of Portland Bill. It is full of oyster fossils, indicating that this limestone was deposited in the sea very close to land. Above left: The Portland coast. Above right: Algal burrs, Kingbarrow Quarry.

Lulworth, Worbarrow Bay and Dorset County Museum

Note: Both the Fossil Forest and Worbarrow Bay are in army ranges and are only open at certain times; usually school holidays and most weekends. For details of opening times check www.dorsetforyou.com/lulworth-range-walks or telephone 01929 404714.

Lulworth is not a particularly good location for fossil hunting but echinoids and molluscs can be found. Look in the rubble on the beach – hammering the cliffs is forbidden. Nearby, however, is one of the gems of the Jurassic Coast, the Fossil Forest. Walk around the cove to the eastern side and climb the path to the top. If the tide is in you will have to take the path on the western side and walk around the top of the cove. Then follow the coast path eastwards and you will soon see the Fossil Forest signposted. On a ledge just above the sea are the petrified remains of algal growths (often called stromatolites) which formed around rotting tree stumps in a coastal swamp about 100 million years ago.

Above and left: Fossilised algal growths in the Fossil Forest.

Dinosaurs would have waded through this swamp and fed on the luxuriant vegetation. The trees were mainly a variety of cypress and were rooted in what is known as the Great Dirt Bed, basically an ancient soil. The trees may have died when the sea level rose slightly giving a more saline environment.

Worbarrow is perhaps not an obvious fossil collecting place, but its charm and proximity to the deserted village of Tyneham make it well worth a visit. Look out for the piece of fossilised wood in the church at

Above: Worbarrow Bay with Worbarow Tout at the far end. This is where to look for dinosaur footprints. The bay is carved out of soft sands and clays from the Cretaceous. Some of these were deposited in a river delta and you can pick out bands of lignite formed from ancient forests.

Tyneham. The cliff at the far end of the bay is Worbarow Tout where you can see Purbeck limestones which were deposited in coastal lagoons. Dinosaurs waded through them and footprints have been found here. Look out for three pronged indentations on the flat bedding surfaces and use your imagination! Also look out for desiccation cracks (right), formed when the lagoons dried out. There are also plenty of bivalves to be found in the Purbeck beds.

Dorset County Museum, Dorchester

The Dorset County Museum in High West Street, Dorchester has a Jurassic Coast Gallery with lots of information about the fossils that are found there and the environments in which they lived. It also has an impressive fossil collection on display, the centrepiece of which is a 2.4 metre long skull of a fossil pliosaur. This creature was probably around 16 metres long and weighed as much as 12 tons. Pliosaurs were the largest marine reptiles that have ever lived and with their crocodile like heads had tremendously powerful jaws with long, sharp teeth. The specimen, which is around 155 million years old, was found by amateur collector Kevan Sheehan in Weymouth Bay and subsequently purchased by the museum.

Kimmeridge

Kimmeridge is famous the world over as a fossil location. The rocks of the Kimmeridge Clay formation were deposited in a warm, shallow Jurassic sea around 150 million years ago. They are mainly soft clays, mudstones and oil shales punctuated by harder bands of limestone. The sea, close to land, was teeming with life and many creatures became fossilised after sinking to its floor.

We must begin with a word of warning – the cliffs at Kimmeridge are extremely unstable, keep away from the base of the cliff. It is not allowed to hammer, and fossils can only be collected from loose material on the beach. Since Kimmeridge is a popular tourist location, finding fossils to collect can prove difficult. You can, however, collect in a different way; many ammonites are visible in the surface of the hard limestone ledges and boulders on the beach. It is not possible or permitted to extract these, but you can photograph them. They make excellent photographic subjects. As well as ammonites, bivalves are common and you may be lucky enough to see bones of marine reptiles in the rocks.

A toll road leads to the sea where there is a large cliff top car park.

The Etches Collection

A Kimmeridge resident, over many years Steve Etches has built up a remarkable and world famous collection of fossils from the Kimmeridge Clay of Dorset. Many of his finds are now on display in a modern, purpose built museum in the village. There is no better place to learn about life in the Jurassic seas.

The Etches Collection Museum of Jurassic Marine Life, Kimmeridge, Dorset, BH20 5PE.
www.theetchescollection.org

Above: Possibly the remains of a bone from a marine reptile. Right: Ammonites are common in the rocks along the beach. They are often fragile, why not "collect" them with a camera.

The Isle of Purbeck

The Isle of Purbeck includes Kimmeridge which is dealt with separately. Limestone is the dominant rock on the coast, the marine Portland limestone is overlain by the Purbeck limestones which were mainly formed in shallow, coastal lagoons. To the north of the region the Chalk ridge runs east to west while in between this and the limestones lie soft sands and clays, mostly deposited in a large river delta. This diversity of rock strata results in a great variety of fossils.

The Portland limestone contains many ammonites and the giant Titanites is one of the zone fossils for this rock. There is a splendid example by the shore in the rocky cove at Winspit and two more lie side by side in the cliff face in the quarry near Anvil Point lighthouse. The Purbeck limestones were largely deposited in coastal lagoons and the fossils reflect this different environment. There are plenty of freshwater snails and also petrified worm burrows. Lookout for desiccation cracks, formed when the bottom of the lagoon periodically dried out and then filled with sediment when water returned (see page 33).

The rocks of Durlston Bay are a classic section through the Purbeck Beds, mostly thin bands of limestone interspersed with softer bands of sands and clays. They date from the top of the Jurassic and bottom of the Cretaceous periods around 140 million years ago and were formed in a coastal lagoon of varying depth. There have been important fossil finds here, including remains of early mammals and, more recently, a large crocodile skull. However, it is not really a place for family collecting – the area is prone to cliff falls and the

Right: Anvil Point, Durlston Country Park. In the small quarry just below the lighthouse two giant ammonites (Titanites) lie flat on what was once the bottom of a tropical sea. This is the Portland limestone, the Purbeck limestones lie on top of it.

*Above: Worm burrows (foreground) in the Portland limestone at Winspit.
Below: Purbeck limestone with dozens of bivalve shells.*

rocky shore is narrow. It can be accessed from Peveril Point on the south side of Swanage, but please take care and stay away from the cliffs.

Some limestone strata are crammed full of bivalves and snails, while fish and crocodile teeth are sometimes seen scattered through the pale limestones – look out for shiny black bits (see photograph page 20). If you plan to go here, first make time to visit the Rock Room at Durlston Castle. This excellent facility has a large slab of Purbeck limestone with all the things you can expect to find, while an informative video presentation explains all about the ancient environments at the time of formation.

Above: Purbeck limestone with many tiny bivalves but also what looks like a shark's tooth (inside box).

The swampy coastal lagoons in which the Purbeck Beds were deposited were also home to dinosaurs, including the giant plant eaters such as Brachiosaurus. These creatures left their footprints in the soft mud of the lagoons and chance circumstances such as rapid desiccation followed perhaps by a rise in sea level and more sedimentation have preserved some of them in the bedding planes. You may be lucky enough to find an example yourself, but to be sure to see some go to Keates Quarry near Worth Matravers. More than one hundred fossilised dinosaur footprints are preserved here; thought to be of giant sauropods like Brachiosaurus. There are even ones presumed to be made by juveniles and a mark thought to have been made by a tail. The tracks were discovered by quarry workers in 1997 and opened to the public only in 2016. The quarry is located near the Priests' Way footpath east of Worth Matravers at GR SY98671 78046.

Above: Dinosaur footprints at Keates quarry; the shoe in the top left picture gives an idea of the size these creatures must have been!

Places to see:
Winspit, near Worth Matravers, Portland limestones with giant ammonite, bivalves and worm burrows.
Square and Compass, Worth Matravers, excellent little fossil museum in this well known pub.
Durlston Country Park, see the Rock Room and Fossil Wall at the Visitor Centre.
Langton Matravers Museum, a charming, small museum with a fossil collection.
Swanage Heritage Centre and Museum, lots of information and a small fossil collection.
Keates Quarry, Worth Matravers, over 100 dinosaur footprints.

The Classification of Animals (Taxonomy)

The classification of animals is a complex business with a multitude of names referring to different groupings. The common names of fossil groups do not necessarily refer to the same sort of group. The table below shows the main divisions of the animal kingdom with, as an example, the names of those that include humans. The divisions are listed in decreasing order of size.

Division	Humans
Phylum	Chordata
Class	Mammalia
Order	Primates
Family	Hominids
Genus	Homo
Species	Homo sapiens

It might be of interest to know where some of the common fossils we find on the Jurassic Coast fit into this classification.

Ammonites are molluscs (phylum) from the class Cephalopoda and the sub-class Ammonoidea.

Bivalves likewise are a class of molluscs, while belemnites are an order of the cephalopods.

"Fish" does not describe any one division but rather includes a number of classes.

Dinosaurs are grouped in the super-order Dinosauria which consists of the two orders Saurischia and Ornithischia.

Finally echinoids are part of the class Echinoidea in the phylum Echinodermata.

Fossil Collecting Code

Before embarking on a fossil collecting expedition please visit jurassiccoast.org/conserving-the-coast/fossil-collecting. Here you will find a comprehensive guide to responsible and safe fossil collecting. A very brief guide is summarised below:

The best, and safest, place to look for fossils is on the beach where the sea has washed away soft clay and mud. Low tide is best.

Do not collect from or hammer into the cliffs, fossil features or rocky ledges.

Keep collecting to a minimum, leave something for others.

Avoid removing in situ fossils.

Never collect from walls or buildings. Take care not to undermine fences, bridges or other structures.

Be considerate and don't leave a site in an unsightly or dangerous condition.

Please observe local notices.